TOO CUTE! BABY ANIMALS

BABY FOREST ANIMALS

RACHAEL MORLOCK

PowerKiDS press

Published in 2025 by The Rosen Publishing Group, Inc.
2544 Clinton Street, Buffalo, NY 14224

First Edition

Editor: Rachael Morlock
Book Design: Michael Flynn

Photo Credits: Cover, p. 1 Steve Byland/Shutterstock.com; (series background) Viktoriia LiSa/Shutterstock.com; p. 5 Ben McMurtray/Shutterstock.com; p. 7 Manamana/Shutterstock.com; p. 9 Wonderly Imaging/Shutterstock.com; p. 11 Japan's Fireworks/Shutterstock.com; p. 13 Harry Collins Photography/Shutterstock.com; p. 15 Debbie Steinhausser/Shutterstock.com; p. 17 Nikolay Zaborskikh/Shutterstock.com; p. 19 FotoRequest/Shutterstock.com; p. 21 Mike Wilhelm/Shutterstock.com.

Library of Congress Cataloging-in-Publication Data

Names: Morlock, Rachael, author.
Title: Baby forest animals / Rachael Morlock.
Description: [Buffalo] : PowerKids Press, [2025] | Series: Too cute! Baby animals | Includes index.
Identifiers: LCCN 2024004117 (print) | LCCN 2024004118 (ebook) | ISBN 9781499445237 (library binding) | ISBN 9781499445220 (paperback) | ISBN 9781499445244 (ebook)
Subjects: LCSH: Forest animals–Infancy–Juvenile literature. | Forest ecology–Juvenile literature.
Classification: LCC QL112 .M69 2025 (print) | LCC QL112 (ebook) | DDC 591.3/92–dc23/eng/20240229
LC record available at https://lccn.loc.gov/2024004117
LC ebook record available at https://lccn.loc.gov/2024004118

Manufactured in the United States of America

CPSIA Compliance Information: Batch #CSPK25. For Further Information contact Rosen Publishing at 1-800-237-9932.

CONTENTS

Tree Lovers

Forests are places where trees outnumber other plants and animals. Many animals live in forests. They make their homes in, around, and under different kinds of trees. The trees offer food and **shelter**. Cute baby animals grow up in forests all around the world!

Sloth Babies

Sloths live in rainforests in Central and South America. A baby sloth holds onto its mother's chest as she hangs on branches and moves through the forest **canopy**. The mother and baby stay together in this tight hug until the baby is about six months old.

Panda Cubs

Giant pandas live in China's bamboo forests. Their babies are born in the fall. Each cub is tiny compared to its mother! She holds her newborn close to her body for the first few weeks. The cub stays in a den for about 100 days.

Koala Joeys

Koalas live in Australian forests. A koala baby is called a joey. A newborn is the size of a jelly bean! The joey lives in a pouch on its mother's body for at least six months. After that, the joey rides on its mother's back.

Fox Kits

Red foxes live on the edge of forests around the world. A mother gives birth to about five kits in an **underground** den. The kits drink their mother's milk and stay safe in the den for five weeks. Their father also brings food to the den.

Fawns

Baby deer are called fawns. They're often born as twins in late spring. White-tailed fawns have a spotted coat. If they're lying down in tall grass or on the forest floor, then their coat works as **camouflage**. Fawns drink their mother's milk for four months.

Moose Calves

Moose live in northern forests in North America, Europe, and Asia. Their tan-colored calves are born in spring, and twins are common. The calves drink their mother's milk right away. For the next year, the calves sleep, swim, and eat plants by their mother's side.

Owlets

The barred owl lives in North American forests. Barred owls often lay their eggs inside tree hollows. Chicks, called owlets, **hatch** in the spring. Owlets are covered in a fluffy white down. They stay in the nest with their mother, and their father brings them food.

All Kinds of Cute

The animals in this book are **mammals** or birds that start life as soft, fuzzy creatures. Forests are also full of baby **reptiles** and **amphibians**. Turtles, lizards, snakes, salamanders, frogs, and toads may not be fluffy, but they can certainly be cute too!

GLOSSARY

amphibian: An animal that lives part of its life in water and part on land.

camouflage: Colors or shapes on something that allows it to blend in with its surroundings.

canopy: The highest layer of branches in a forest.

hatch: To break open or come out of.

mammal: Any warm-blooded animal whose babies drink milk and whose body is covered with hair or fur.

reptile: A cold-blooded animal with thin, dry pieces of skin called scales.

shelter: A place where animals or people are kept safe.

underground: Below the surface of the earth.

FOR MORE INFORMATION

BOOKS

Jenner, Caryn. *Rainforest Animals*. New York, NY: DK Publishing, 2021.

Neuenfeldt, Elizabeth. *Forest Animals*. Minneapolis, MN: Bellwether Media, 2023.

WEBSITES

Barred Owl

www.chattnaturecenter.org/visit/experience/wildlife/animal-facts/barred-owl/

Read up on the barred owl, which stays with its young longer than most other owl species.

Koala

kids.nationalgeographic.com/animals/mammals/facts/koala

Find fascinating facts about koala joeys—and what they eat—on this website.

Publisher's note to educators and parents: Our editors have carefully reviewed these websites to ensure that they are suitable for students. Many websites change frequently, however, and we cannot guarantee that a site's future contents will continue to meet our high standards of quality and educational value. Be advised that students should be closely supervised whenever they access the internet.

INDEX